EARTH IN DANGER

HABITAT DESTRUCTION

Words that appear in **BOLD** are explained in the glossary.

First published by Hungry Tomato Ltd in 2022
F1, Old Bakery Studios, Blewetts wharf,
Malpas Road, Truro, Cornwall, Tr1 1QH, UK.

Copyright © Hungry Tomato Ltd 2022

www.hungrytomato.com

ISBN 978 1 914087 88 2

A CIP catalogue record for this book is available from the British Library.

Printed in China.

Picture credits (t=top, b=bottom, m=middle, l-left, r=right, bg=background)

IStock: 18br; 20-21bg. **Shutterstock:** 10bl; 17br; 21br; 24br; 4-5bg; 4bl; 6bl; Al'Fred 18-19bg, 30br; Aleksandr Rybalko 10-11bg; Amino 9br; Anouska13 28mr; Anton Gvozdikov 6-7bg; Avigator Fortuner 2bg; Berhard Staehli 25mr; Blue-sea.cz 27tr; e2dan 22-23bg; Edward Haylan 32br; GUDKOV ANDREY 25bl; In Green 26b; Inside Creative House 29ml; Jay Ondreicka 1bg; Jeremie Thomas 29b; Jirapa339 24ml; Keat Eung 31bg; Korotin photographer 29tr; L.F 13tl; Rawpixel.com 28bl; Stockr 8-9bg, 32mr; Subphoto.com 27ml; Symbiosis Australia 14-15bg; Thammanoon Khamchalee 12-13bg; The Len 27br; Tom Payne 3bg; Viktoriya Krayn 16-17bg.

Every effort has been made to trace the copyright holders, and we apologise in advance for any unintentional omissions. We would be pleased to insert the appropriate acknowledgements in any subsequent edition of this publication.

CONTENTS

Just think of all the different places in the world where wild plants and animals live. From your garden or local park, to tropical **rainforests** and the freezing polar regions, our planet is covered in amazing **habitats.**

A habitat is a place where an animal or plant lives. Each animal and plant has **adapted** to survive in its habitat.

So, if a habitat is destroyed, wildlife can't just move somewhere else. Habitat destruction is putting many animals and plants in danger of becoming extinct.

Ring-tailed lemurs only live in the forests of Madagascar. Their habitat is in danger.

AMAZING HABITATS

Our planet's population is increasing. In 1950 there were two and a half billion people in the world. By 2050, there could be as many as 9.9 billion people!

This growth means that we need more food, water, clothes and places to live. We also need materials like stone, wood and metal for building, and fuels to power vehicles, homes and businesses.

In order to supply all of these things, many natural habitats are being destroyed. This means fewer places for animals and plants to live.

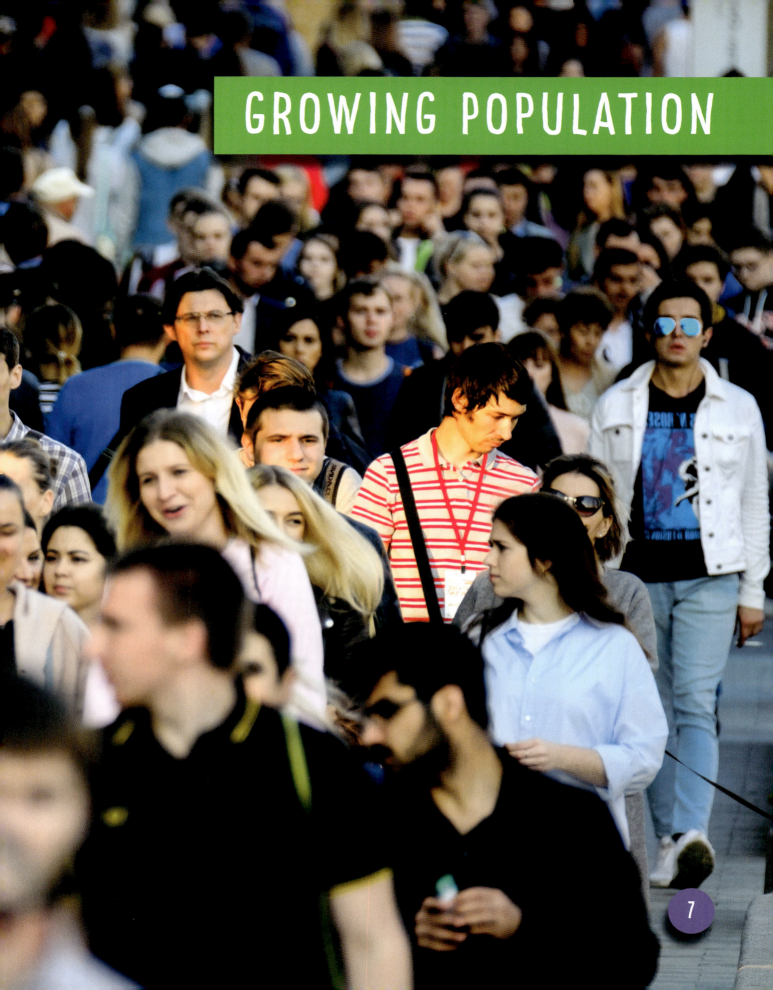

GROWING POPULATION

LAND FOR CROPS

DID YOU KNOW?

As Earth's population continues to grow, more farms will need to be created. That could mean more habitats are destroyed.

In order to feed Earth's growing population, more food needs to be produced. However, there is not enough **fertile** soil to do this.

For hundreds of years, many forests, grasslands, and other natural habitats, have been cut down or cleared to make room for growing crops and grazing animals.

We need to find a way to save these wild habitats and also feed everybody.

To feed everyone, farmers don't just need more land to grow things, they also need to grow more food on the land they have. This is called **intensive farming.**

Intensive farming means using more machines to grow more crops.

Pesticides are used to get rid of weeds and insects that damage crops. These pesticides can be harmful to habitats, as they can poison other plants and animals that live nearby.

DID YOU KNOW?

Intensive farming may also use chemical fertilisers to make crops grow bigger and faster. These can pollute nearby water habitats, killing wildlife.

Chemicals that get into nearby water can kill fish.

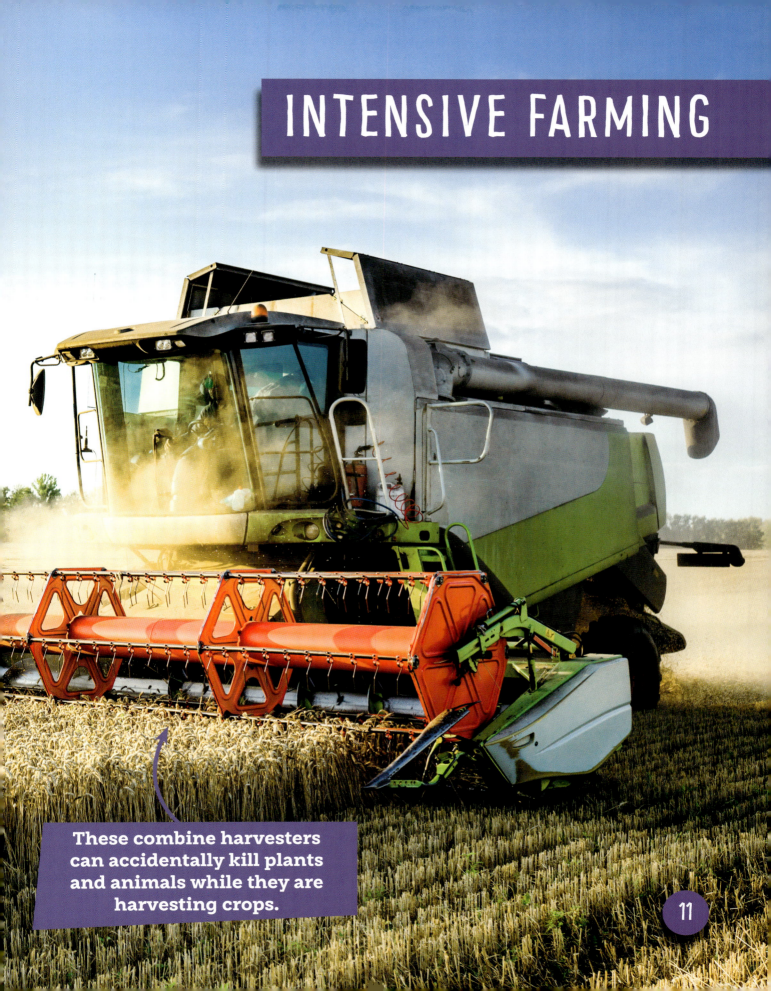

INTENSIVE FARMING

These combine harvesters can accidentally kill plants and animals while they are harvesting crops.

RAINFOREST TIMBER

DID YOU KNOW?

Forests cover around 30% of the land on our planet, but it used to be much more!

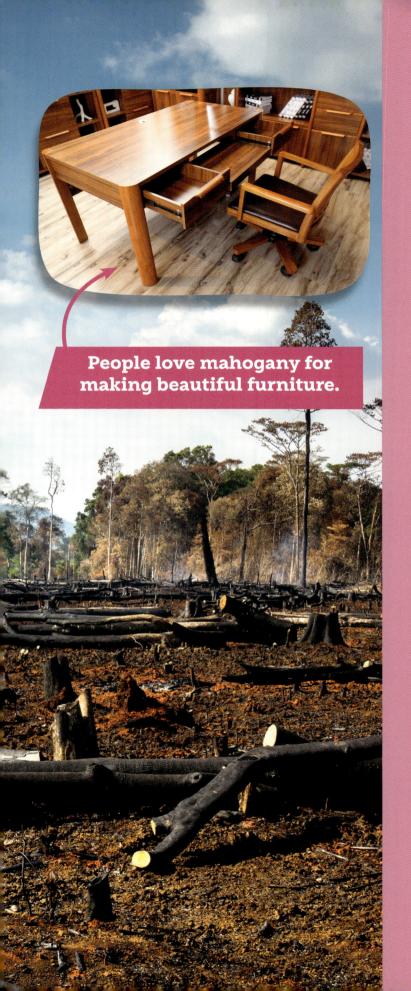

People love mahogany for making beautiful furniture.

The increase in human population also means more homes are needed. Many trees used for building homes and furniture come from **sustainable** forests.

This means more trees are planted to replace the ones that are cut down. This can still cause damage to habitats, as forests can take a long time to grow back.

Not all timber is sustainable. Rainforest timber, like **mahogany**, is valuable because it is strong and attractive. Once these trees are cut down, the land is usually cleared to use as farmland. This means that the original wildlife habitat is lost.

Many useful or valuable materials come from the ground. These include metals, such as iron, copper or gold, and fuels, like oil, gas and coal.

To reach the materials, they have to be mined from the ground. Many of them are found underneath wildlife habitats.

It is impossible to mine an area without damaging the habitat above. Heavy machinery is used to remove rock and soil, and **extract** the valuable materials. Roads are built to transport the mined materials.

These changes to the land can kill or drive out the animals and plants that live there.

DID YOU KNOW?

Areas of water near to mines can easily become polluted. This can also harm wildlife.

DAMAGE FROM MINING

Open cut gold mining operation

USING TOO MUCH WATER

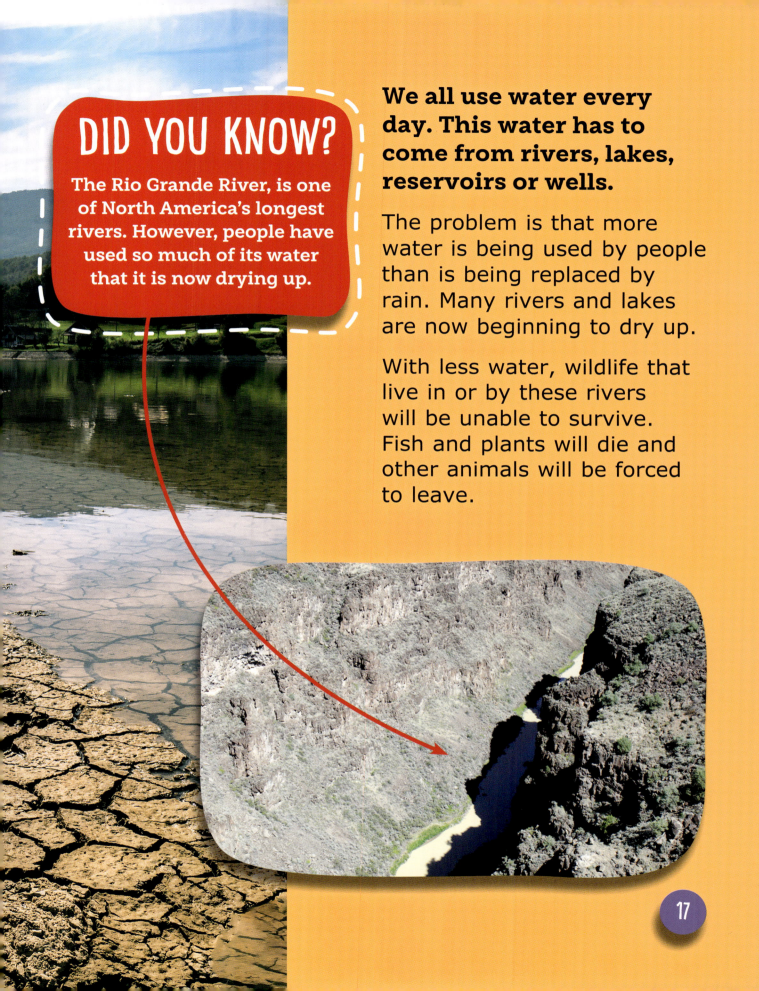

We all use water every day. This water has to come from rivers, lakes, reservoirs or wells.

The problem is that more water is being used by people than is being replaced by rain. Many rivers and lakes are now beginning to dry up.

With less water, wildlife that live in or by these rivers will be unable to survive. Fish and plants will die and other animals will be forced to leave.

Rainforests are home to the greatest number of different types of plants and animals on Earth. Many of these are very useful to us.

Deforestation continues to destroy large parts of these precious habitats.

Too much of the gas, **carbon dioxide**, in the Earth's **atmosphere** is causing the planet to become too warm. We call this **global warming**, and it could have serious effects on habitats around the world.

Rainforest plants can help by **absorbing** carbon dioxide. If we keep destroying rainforests, we will be putting many other habitats, as well as our own, in danger.

DID YOU KNOW?

Rubber is made by cutting into a rubber tree and collecting the sap.

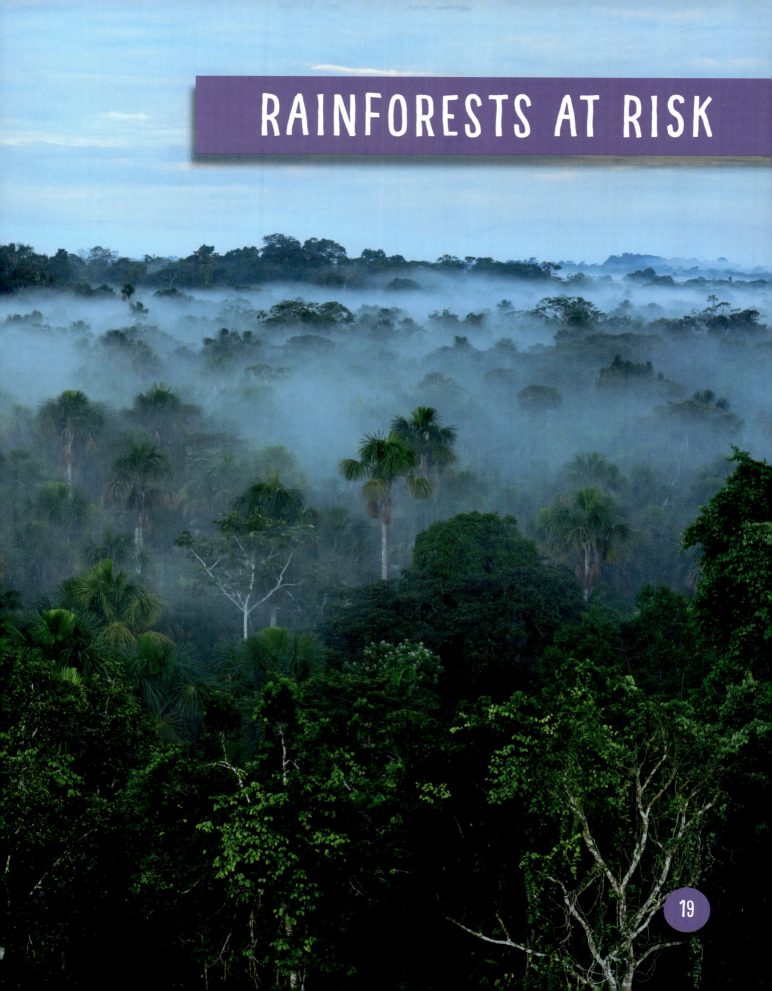

RAINFORESTS AT RISK

CLIMATE CHANGE & HABITATS

DID YOU KNOW?

Climate change is making wildfires more common. Fires can easily destroy wildlife and their habitats.

Climate change, in the form of global warming, can be a big cause of habitat destruction. Changes to the climate are happening faster than ever before.

These changes include rising temperatures and more extreme weather, such as powerful storms and **droughts**.

Rising temperatures and droughts can dry out the land, including rainforests and swamps. This kills plants and leaves animals with very little food or water.

Extreme flooding, caused by storms and rising sea levels, can also be very damaging to natural habitats.

Hotter temperatures around the world may melt ice at the North and South Poles.

PROTECTING THE PLANET

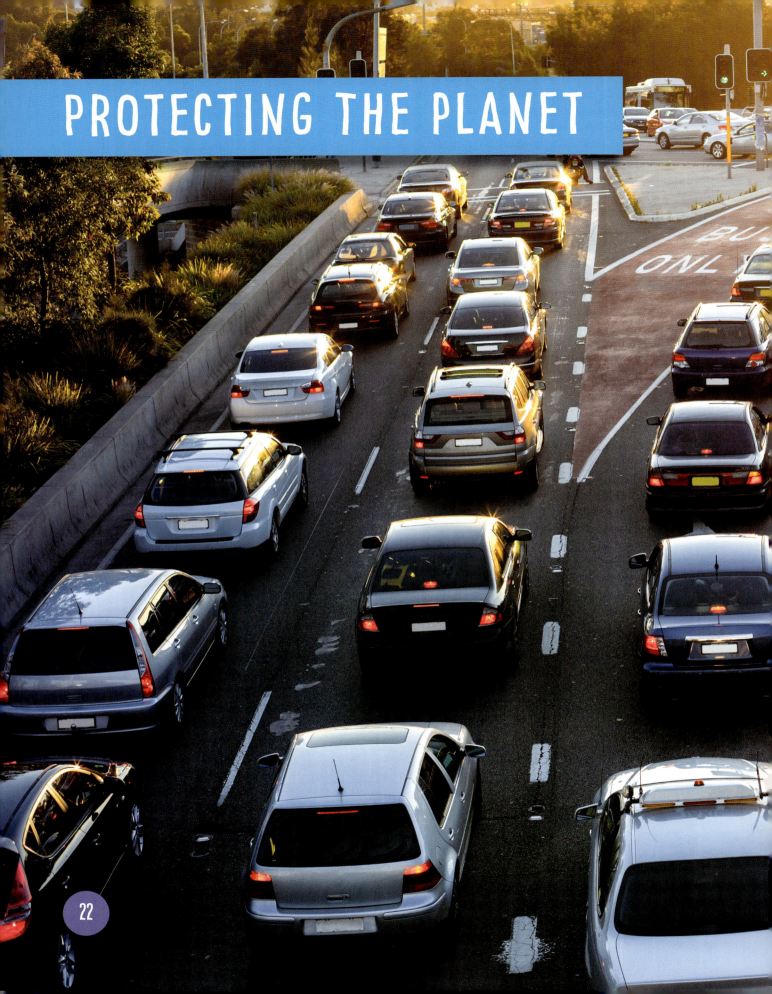

Two big causes of habitat destruction are population growth and climate change. These two causes are connected.

The more people there are, the more vehicles and machines we need to use. These things need power to keep them running. Most of that power comes from fuels that release carbon dioxide into the atmosphere.

As we clear forests and rainforests to produce more food, materials, and space for buildings, we increase the levels of carbon dioxide, as there are less trees and plants to help absorb it.

To stop habitat destruction, we need to protect the planet from damage caused by both population growth and climate change.

RAINFOREST FACTS

Here are some rainforest facts to give an idea of how important this type of habitat is:

- There are so many different animals and plants in rainforests that scientists believe there are some that are yet to be discovered.

- If we continue to destroy rainforests, some **species** could become extinct before we have even discovered them!

The Rosy periwinkle, found in Madagascar, is used in some cancer-fighting drugs.

- Many rainforest plants are used to make medicines.

24

POLAR REGION FACTS

Polar habitats, in the Arctic and Antarctic are also in danger from climate change.

- Global warming is causing sea ice in the Arctic to melt.

- The melting ice is causing our sea levels to rise. This could lead to flooding in different parts of the world.

- Animal species, like polar bears and some types of seal, need this sea ice to be able to hunt and raise their young.

CORAL REEFS

Coral reefs, found in the ocean, are one of the world's richest and most beautiful habitats. Here are some facts about them:

- Coral is made from the skeletons of tiny animals called polyps. When these polyps group together, they form large structures called reefs.

DID YOU KNOW?

The Great Barrier Reef in Australia is so big, it can be seen from space!

- Coral reefs provide a home for thousands of different species of plants and animals.

Plastic rubbish on a coral reef

- Coral reefs are in danger. They can be damaged by pollution, warming temperatures, overfishing, and physical destruction by people and ships.

- Artifical (human-created) reefs are being made to replace coral reefs that have been destroyed. They provide new homes for the animals and plants.

SAVING LOCAL HABITATS

Look around your neighborhood. There are different habitats all around you! Here are some ways to help support them:

- Plant some flowers! Nectar in flowers provides food for bees, butterflies and other insects, and they are very pretty to look at too!

A bug hotel

- Create log piles or build a bug hotel. These make great places for insects and other small animals to live.

- Ponds are great habitats for lots of species. Why not try making a mini pond from a recycled waterproof container? Instructions are easy to find online.

- Rubbish that is left scattered around can be very harmful to wildlife. Always throw your litter away, and recycle it if possible.

HOW YOU CAN HELP

Here are some ways you can help to save habitats around the world:

ENERGY SAVER!

Using less electricity can help reduce global warming. Try:

- Turning off lights and electronics when you aren't using them. Don't leave them on stanby.
- Walk or cycle instead of using a car.
- Take holidays closer to home to avoid using planes.

REDUCE, REUSE, RECYCLE

- Reusing materials saves energy and cuts down on waste, which causes pollution and harms wildlife. Using recycled paper and wood products means less trees have to be cut down.

adapted Changed or suited to fit into a particular situation or habitat.

atmosphere The air, and gases that surround our planet.

carbon dioxide A gas given off when things decay or are burnt.

climate Patterns of weather over a long period of time.

droughts A long period of time with little or no rainfall, leading to a shortage of water.

deforestation Cutting down large areas of trees or forest.

extinct A species of animal or plant that has disappeared forever.

extract To remove.

fertile Rich in substances that help plants grow; capable of growing in great numbers or quantity.

global warming The warming of the planet's air and oceans as a result of a build-up of greenhouse gases in the atmosphere.

habitat The place that suits a particular animal or plant in the wild.

intensive farming Using money, equipment, and chemicals to get more crops and other food products out of less land.

mahogany A heavy, strong wood, often used to make furniture.

mined Removed from deep inside the Earth.

overfishing Taking more fish from a body of water than can be replaced naturally.

oxygen A gas in the Earth's atmosphere. Humans and most other living things need oxygen to breathe.

pesticides Chemicals that kill animals or plants that damage crops.

population The number of people, or animals living in a certain place.

pollute To release harmful substances into the environment.

rainforest Huge forests of tall trees. Rainforests are normally warm and have lots of rain.

sustainable Something that can be used into the future without using up resources, or made in a way that will last forever.

species A specific type of plant or animal.

31

INDEX